Nandika

Omar

Dorian

D1636904

Nneka

Martin

Kaichang

3

What Do You Want to Be?

Is working in green chemistry one of your goals?

The good news is that there are many different paths leading there. The people who work in green chemistry come from many different backgrounds. They include chemists, engineers, environmental scientists, writers, teachers, nanoscientists, environmental activists, and more.

It's never too soon to think about what you want to be. You probably have lots of things that you like to do—maybe you like doing experiments or drawing pictures. Or maybe you like working with numbers or writing stories.

Cool Careers in
GREEN CHEMISTRY

Sally Ride
Science

CONTENTS

Meg

Paul

Jim

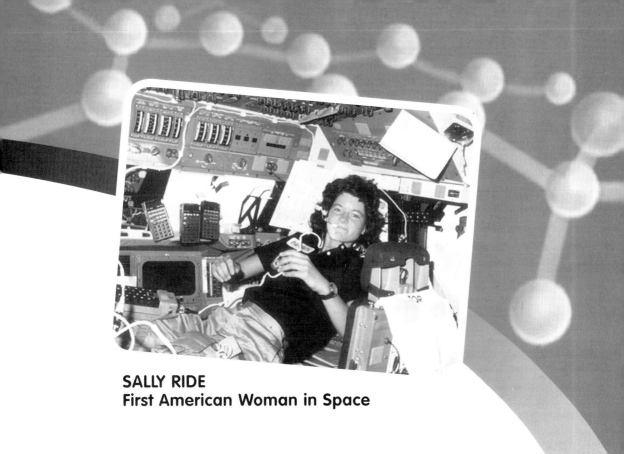

SALLY RIDE
First American Woman in Space

The women and men you're about to meet found their careers by doing what they love. As you read this book and do the activities, think about what you like doing. Then follow your interests, and see where they take you. You just might find your career, too.

Reach for the stars!

Sally K Ride

JOAN NIERTIT

South Coast Air Quality Management District

Job Ladder

Joan Niertit used to climb smokestacks to collect samples of pollution belching into the atmosphere. The tallest was 60 meters (200 feet)! Joan is now back to Earth, but she keeps her eye on the sky. Her focus? Smog—the nasty stuff formed when sunlight reacts with air pollutants such as hydrocarbons. One source are the solvents needed to make paint spreadable. Drying paint can release hydrocarbon fumes for months. So Joan tests paints to make sure they meet strict environmental limits. "The limits force companies to come up with cleaner and cleaner paints," Joan says. Breathe easy—Joan is snuffing out another source of smog!

No Complaint Paint

Joan's work keeps paint companies busy finding green alternatives to oil-based solvents —such as good old water. Even though the "green way" usually ends up saving companies money, Joan says many still need a push. But many companies have chosen to go green. Some paints contain just one-tenth the polluting solvents they did just decades ago. That means less smog *and* fewer people with lung ailments—just ask Joan. "As the smog has gotten better, so has my asthma," she says.

Breath of Relief

Hate being stuck indoors? Good thing you didn't live in Los Angeles in the 1970s! Back then, heavy smog made it unsafe to play outside about 100 days a year. Today, it's still the smoggiest place in America. At least now it's safe to run, bike, and skate outdoors without seriously hurting your lungs.

Hold your breath! Joan checks one of the canisters used to capture samples of polluted air.

An air quality chemist

analyzes, and works to eliminate, contaminants that pollute the air. Joan researches how to reduce the smog-forming fumes from drying paint. Other **air quality chemists**

* measure exhaust from car and truck tailpipes.
* help design cleaner coal-burning power plants.
* pinpoint sources of harmful fumes.

Can U C Your Air?

How much particle pollution, such as soot and dust, is in the air you breathe? As a class, make a hypothesis then try this five-day experiment.

- Use a jar lid to trace five circles on white poster board. Label the circles and five lids 1-5.
- Put the poster board outside, then place each lid over its matching circle.
- At the end of each day, remove one lid and compare the circles.
- Record your observations.
- After you remove lid 5, compare all the circles. Use a magnifying glass to count particles.

Which circles have the most particles, and why? Is your air clean or dirty?

Is It 4 U?

What parts of Joan's job would you like?

- Climbing smokestacks to collect air samples
- Testing paints to see if they meet environmental standards
- Using chemistry to analyze air samples

Team up with a classmate and take turns role-playing Joan. Explain what you like best about your job.

Bad Air Day

That's what you call it when the air where you live is smoggy, sooty, or otherwise polluted.

- As a class, track the Environmental Protection Agency's Air Quality Index in your town every day for one month.
- Create a bar graph to show how many days were rated good, moderate, unhealthy, very unhealthy, or hazardous.
- How does your town measure up?

> "You quickly realize it's more intelligent not to produce the waste in the first place."

Right from the Start

Thousands of factories in northern Mexico churn out TVs, clothes, car parts—and toxic chemical waste. Yuck! One of Conchita's early jobs in her native Mexico was figuring out how to clean it up. That's why she now designs chemical processes that are "green" from the get-go. "It's all about getting it right the first time," she says.

CONCEPCIÓN JIMÉNEZ-GONZÁLEZ
GlaxoSmithKline

Green Medicine

Concepción "Conchita" Jiménez-González works for a company that makes medicines—everything from vaccines to asthma inhalers—that make people better. Conchita's job is to make those medicines better . . . for the environment. That's because it can take some bad chemicals—such as dichloromethane and toluene—to create good medicines. Those types of chemicals are expensive to clean up—and to buy and to use, too. "You pay for them three times," she says.

Waste Not

And get this—making 1 kilogram (2.2 pounds) of medicine can generate as much as 90 kilograms (about 200 pounds) of leftover chemical waste. What a pill! So Conchita researches how using more eco-friendly chemicals, in smaller amounts, can achieve the same results—without all the nasty waste. There's often more than one way to make a medicine. Conchita creates guides that help her fellow chemists pick ways that are least harmful for the environment. She works with the chemists to help them select methods that are safer, faster, and cheaper. Sharing how she learned to ride a bike at age 35 can be a big encouragement!

Conchita's love for dancing started when she was young—and she's still moving to the beat.

A chemical engineer

uses chemistry and engineering to develop safe and inexpensive ways to use chemicals. Conchita studies ways to make greener medicines. Other **chemical engineers**

* develop nontoxic plastics for kitchen use.
* design exhaust systems that cut tailpipe pollution.
* create batteries that power electric cars.

Gray into Green

Don't waste that water! "Gray water"—water that comes from the tub, shower, bathroom sink, and washing machine—is recyclable. After filtering, it's safe for watering grass, shrubs, and trees outdoors. "Black water" from the kitchen sink, toilet, and dishwasher is too dirty and dangerous to use.

How much gray water does the average person produce?

* Shower and tub—95 liters (25 gallons) per person per day
* Washing machine—60 liters (16 gallons) per person per day
* Bathroom sink—8 liters (2 gallons) per person per day

How much gray water does your family produce in a day? In a week? In a year?

All Set

When Conchita was young, her favorite plaything was her chemistry set. What would her chemistry set look like today? In addition to the usual tools such as goggles, beakers, funnels, and tongs, there might be green chemicals, such as these.

Table salt	Glycerin
Cornstarch	Alum
Vinegar	Rock salt
Unflavored gelatin	Baking soda
Sugar	Mineral oil
Flour	Lemon juice

Do some research and find a green chemistry experiment. As a class, share your experiments and assemble a book, *Green Chemistry Investigations*.

Throw Out the Garbage

Why not start making things better for the environment with a no-garbage day? That means not creating any waste—or if you must, reusing or recycling it. Write an About Me Journal entry describing both the challenges and the rewards.

JENNIFER KING
The Community School

Class Power

When Jennifer King teaches chemistry, she fuels the imagination of her students—and a car or two. Jen's students make their own fuel—Earth-friendly biofuel. Their source isn't oil drilled from the ground. It's leftover vegetable oil that a local restaurant used to fry food. Jen's students use a chemical process called esterification to turn that waste oil into biofuel. Way to go! That's just one way Jen teaches her students green chemistry—chemistry techniques that are good for the environment.

Down the Drain

Chemistry wasn't always green at Jen's school. When she started teaching, the school spent thousands of dollars to safely empty a storehouse of old and dangerous chemicals. Now, Jen tries to use only nontoxic household chemicals whenever possible in her classroom. "I prefer using materials that are safe enough to go down the drain," Jen says.

Now We're "Tocking"

In one experiment, called a chemical clock reaction, Jen's students use household supplies to investigate a chemical reaction. They add vitamin C and iodine to distilled water in one beaker, then add laundry starch and dilute hydrogen peroxide to distilled water in another. Both solutions are clear—until combined. *Tick-tock*. Start the clock! After about 30 seconds, the chemicals react and the solution turns dark blue. Jen's students then adjust the temperature or amount of water to speed up—or slow down—the reaction time. It's fun chemistry that's safe, cheap, and easy to clean up.

Jen and her students catch their breath after a hike that's part of their yearly class campout.

A chemistry teacher instructs students about chemical elements, compounds, and reactions. Jen includes green chemistry in her lessons. Other **chemistry teachers** teach

✴ how chemicals cycle through living things and the environment.

✴ ways to replace toxic chemicals with harmless chemicals.

✴ how to design experiments that use less energy.

Whoosh! **Green chemistry makes cleanup faster. That's really great in winter when Jen and her students ski almost every afternoon.**

Fuel for Thought

Even though you don't own a car, maybe you can still reduce your gas consumption.

- Track how many miles you travel by car or bus for one week.
- Find out the miles per gallon for the car or bus.
- Divide by the average number of people aboard.
- Then calculate how many gallons of gas each person uses.

How could you reduce that amount? Brainstorm ideas with a classmate. Then compare your ideas with another team.

Looking for Answers

"I got interested in science because I wanted to understand how the world works," Jen says. Discuss with a partner something you'd like to understand better.

Food for Thought

"Biofuel" is an accurate but boring name for something so cool. Brainstorm with another student a fun name for biofuel made from cooking oil. Here are some of the types of oil that might inspire you.

- Soybean
- Olive
- Palm
- Peanut
- Corn
- Cottonseed
- Canola
- Sunflower

Pass the Ketchup?

If you've ever been around a car that uses biofuel, you might get hungry. Why? You guessed it—the exhaust smells like french fries!

MEG SCHWARZMAN
University of California, Berkeley

Sort It Out

Yikes! Atrazine, chlorobenzene, perchloroethylene . . . the world of chemicals is a soup of strange names. And we're in it! Chemicals are everywhere—the ones above can show up in our food, clothes, and drinking water. They are just some of the more than *80,000* different chemicals we make. So how do we know which ones might be bad for us or the environment? Meg Schwarzman is working on it!

Chemical Caution

"Unfortunately we just don't know the risks of most chemicals," Meg says. Some chemicals may cause immediate harm—and others only after years of exposure. "We might be causing damage we don't even understand yet," she says. So Meg helped write a report about California's need for one set of clear rules to guide how chemicals are made and used. Meg hopes the state will use the report to encourage companies and consumers to make smarter, greener chemical choices. That will mean learning how a chemical could affect people and the environment *before* we make or use it. In turn, that will make it easier to find greener substitutes. Good work, Meg!

Into Helping Out

While training as a medical doctor, Meg treated kids with asthma at a San Francisco clinic. Medicines and inhalers helped some. But they didn't address the effects of neighborhood pollution, which often makes kids' asthma worse. That inspired Meg to focus on public health. Now Meg doesn't just treat patients—she helps whole communities.

Meg loves the outdoors. Here she's stretching after a hike in Yosemite.

An environmental health expert investigates, protects, and improves the health of communities. Meg researches how new chemical regulations protect California's people and environment. Other **environmental health experts**

* monitor air pollution and pollen levels in cities.
* measure noise levels in factories to protect workers from hearing loss.
* prevent farm runoff from polluting sources of drinking water.
* design safety gear for workers who handle hazardous wastes.

Healthy Career Choices

Team up with a partner and choose an environmental health job listed above that interests you. Find out more about it. Then write a career description including

* the focus of the career you chose.
* why you chose this career.
* three reasons why this career is important.

Present your career to the class.

Piece of the Pie

When Meg was just a toddler, her family took a two-year sailing trip. If you could take a two-year vacation, where would you go and what would you do? Survey your classmates and their vacation choices. Then classify each vacation into one of these categories.

1. Exploring new places
2. Investigating history
3. Being entertained
4. Learning about other cultures
5. Relaxing and playing

On a data sheet, record how many students chose each category. Then calculate what percentage of your class chose each one. Create a pie chart to represent the breakdown.

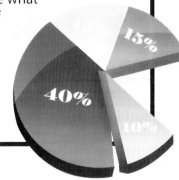

Light Reading

Q. Did you hear about the chemist who was reading a book about helium?

A. She just couldn't put it down!

PAUL ANASTAS
Yale University

Care to Learn

Paul Anastas grew up outside Boston, on the edge of a salt marsh. He loved playing in the reeds, watching ducks and herons, and catching crabs. Unfortunately, developers later bulldozed the marsh—replacing it with office buildings. The loss saddened Paul. His dad's advice? "He told me if I wanted to protect things, I had to learn about them," Paul says. So he did. Paul studied chemistry and the environment. He learned that the way we make, use, and dispose of chemicals often creates a lot of dangerous waste. Waste can be as bad for the environment as a bulldozer!

"There's not an atom of doubt in my mind that green chemistry will ultimately affect everything we see, touch, and feel," Paul says.

Waste? Not!

Paul wondered why chemistry couldn't be helpful to both people and the environment at the same time. So he got to work! Paul pioneered ways to develop chemical products and processes that neither used nor created hazardous materials. Clever. Paul called the new approach "green chemistry." Green chemistry now inspires people worldwide to consider and control the environmental impact of the chemicals they use. This reduces accidents, cuts waste, conserves water and energy, eliminates cleanups—and saves cash.

It Pays to Be Green

In giving green chemistry its name, Paul considered that "Green is not just the color of the environment—it's also the color of money," he says. "Having a healthy environment *and* healthy businesses at the same time, that's what this is all about."

Paul and his Siberian husky, Raven, enjoy the green around them.

A green chemistry professor researches and teaches new and less harmful ways of making chemicals. Paul works with companies, governments, and other experts on saving money while reducing the hazards involved with chemistry. Other **green chemistry professors**

* create plastics from vegetable by-products.
* distill biofuels from renewable plant sources.
* design chemical processes that produce less hazardous waste.
* replace existing products with less toxic alternatives.

Paul has published many, many articles and books, including one of the most widely read books on green chemistry.

Is It 4 U?

What parts of Paul's job would you like? Share your thoughts with a classmate.

- Creating guidelines for a new science
- Flying around the world to give lectures
- Working with students and other professors
- Writing books
- Researching new ways to reduce hazards to people and our planet

Flash Forward

You've probably had an idea for a new invention, new kind of technology, or something you would like to research. Write a newspaper article to announce and describe your contribution to the world of science.

- Begin with a sentence that grabs the reader's attention.
- Write the body of the article, using details to describe your invention.
- Conclude by summarizing all the major points—especially how your idea could make a difference to the world.

Ms. or Mr. History

"When I was young, I liked history a lot. I liked reading about the great people who changed the world," Paul says. Write an entry for your About Me Journal. Include who you like reading about, and why you think she or he is great.

JIM HUTCHISON
University of Oregon

Small but Safe

"When I think about the chemical world," says Jim Hutchison, "I actually imagine being inside it, standing there with the molecules." That's a good thing—it helps Jim keep a close eye on the ultra-microscopic world of nanotechnology. It's the science of creating tiny materials and devices on the scale of individual atoms or molecules. Jim and other scientists are learning how to make new and different nanomaterials—some are even made from gold. These materials are finding their way into everything from medicines to solar panels to computer chips.

Better Safe Than Sorry

There are lots of ways that tiny nanomaterials could become big polluters. So Jim has become a leader in applying the lessons of green chemistry to nanotechnology. That means creating nanomaterials using safer, cheaper, and less wasteful chemical processes. That way, nanomaterials can be made green from the start—cutting out the need for costly cleanups later on. "This is an opportunity to get it right the first time," Jim says.

> "The nanoworld is so small, you can never really see it—so you have to imagine it."

Turn of Fortune

In college, Jim struggled with chemistry until he got serious and started studying more. Soon his future became brighter—not only did he start earning all A's, but his study buddy became his wife!

> Jim volunteers as a ski patroller and teaches a course on the chemistry of snow crystals. Cool.

A green nanoscientist

creates materials and structures on the scale of atoms and molecules. Jim finds new ways to design, make, and use nanomaterials without polluting. Other **green nanoscientists**

* create nanomembranes that filter polluted water.
* engineer nanomaterials used in energy-efficient lights.
* manufacture nanoparticles without using toxic solvents.
* invent self-cleaning nanocoatings for glass windows.

Student Teacher

When Jim was in sixth grade, his favorite science activity was Egg Drop. "It got me really excited," he says. What science activity gets you excited? Get permission to guide another student or another science class through the experience.

- List materials needed.
- Write step-by-step directions.
- Guide the student through the scientific process—ask a question, form a hypothesis, test it, analyze the data, draw a conclusion, share your results.
- Include any cautions, such as wearing safety goggles.

Share your science smarts!

Tiny Invisible World

Jim says the nanoworld is so small, you can never really see it—so you have to imagine it. How well can you imagine it?

Warning—work with a teacher nearby and be careful when using scissors.

- In a shoe box, cut a hole in the end large enough to fit your hand.
- Lay paper in the box, put the lid on, and slide your hand into the box.
- With crayons or markers, draw the water and carbon dioxide molecules below. Label each with its name and chemical formula.

Water (H_2O)   Carbon dioxide (CO_2)

Share your drawings and your nanoworld experience with a classmate.

Nano Laugho

Q. How do scientists count molecules?

A. They atom up.

Nandika D'Souza
University of North Texas

A Dream Come True

Nandika D'Souza is a daydreamer. "People might think daydreamers are not focused, but they're actually doing something creative," says Nandika. The proof? Nandika is dreaming up how to make new environmentally friendly plastics. And, she's using her engineering know-how to make her idea grow—literally!

Planting Plastic

Nandika uses jute, hemp, and a fast-growing plant that's like bamboo called kenaf to make plastics. Plant material can replace the fossil fuel traditionally used to make the plastic in water bottles and sandwich wrap. The advantage? Bioplastics don't take hundreds of years to decompose, like plastics made from petroleum do. One of Nandika's projects is to create plant-based coatings for the paper packaging used in military ready-to-eat meals. The new coating protects the food inside—and the environment outside. That's because it breaks down once the packaging is tossed—eliminating 14,000 tons of waste a year!

Scientifically Literate

"I wanted to be a fiction writer when I was young. I actually had no interest in science or engineering," Nandika says. "In fact, in college, after a year of engineering, I thought about returning to literature. Then I realized that with engineering, I could make things I was daydreaming about."

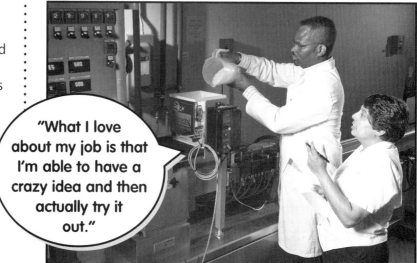

"What I love about my job is that I'm able to have a crazy idea and then actually try it out."

A materials engineer

uses new metals, ceramics, plastics, composites, and other materials to make aircraft wings, tennis rackets, and much more. Nandika creates plastics from renewable plant sources. Other **materials engineers**

✳ engineer plastics to decompose in sunlight.

✳ create lighter and stronger metals used in satellites.

✳ design new ceramics used in artificial hips and knees.

Fibers from the stem of the kenaf plant are used in making everything from textiles to kitty litter.

New and Improved

Good products start with a good design and good materials. Take your bicycle helmet. It needs to cover your head without blocking your view and to be strong enough to protect you. Choose a product from one of these categories—sports equipment, school supplies, gardening tools, or pet stuff. With a partner, brainstorm how to improve your product. Consider three criteria.

- Function—How could it do its job better?
- Design—What specifics would you change?
- Characteristics—Would you change the materials it's made of? Why?

Describe and draw your improved product. Now you're thinking like a materials engineer!

Buy Bioplastics

Good news! From water bottles to mobile phones, bioplastics are now used to make lots of things. Bioplastics are made from renewable plant compounds. Plus, they break down in months instead of hundreds of years. Goodbye, smelly landfills!

With a partner, create a magazine advertisement that encourages people to buy bioplastics. As a class, vote on the most effective advertisement.

Science Qualifications?

Nandika is a daydreamer. Brainstorm with a classmate why you think daydreaming and imagination are important in science.

OMAR YAGHI

University of California, Los Angeles

Toying Around

Omar Yaghi uses the Tinkertoys of chemistry—atoms and molecules—to shape the future. Omar invented new types of material called metal-organic frameworks. The 3-D structures are like microscopic sponges. They're great at soaking up gases such as carbon dioxide and methane. "We can condense the gases into the holes—just like bees in a honeycomb," Omar says. One day, that could let power plants trap greenhouse gases. And, it could let natural gas-powered buses hold twice as much fuel in their tanks. Omar is showing how big changes can come in tiny packages!

Straight to Class

Omar got hooked on the structure, or shape, of chemical compounds while growing up in Jordan. "As soon as I saw the first chemical structure in a book, I fell in love," Omar says. He headed straight to the U.S. to study chemistry. "When I started, I was only interested in the beauty of molecules—not in solving huge societal problems," Omar says.

"I started small, did what I really enjoyed, and only later found success."

Pop Sci

Omar and his colleagues have created hundreds of microscopic structures shaped like pyramids, pinwheels, honeycombs, and stars. They're complicated in shape but cheap to make, "It just takes soda pop bottles and sunblock!" Omar jokes. Well, almost . . . they do rely on common compounds such as terephthalic acid that's used to make plastic bottles, and zinc oxide that's found in many sunscreens!

The microscopic structures Omar builds look like children's toys, but they're serious science.

A materials scientist

discovers and develops new compounds and materials. Omar designs and builds microscopic chemical structures useful in improving transportation and fighting pollution. Other **materials scientists**

* develop cleaner ways of refining oil.

* design easy-to-recycle plastics.

* reformulate paints to remove harmful solvents.

* invent lighter but stronger synthetic fibers.

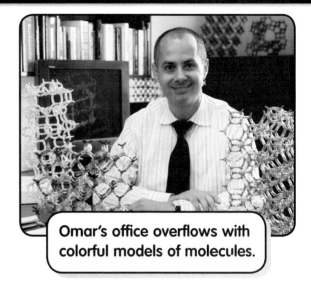

Omar's office overflows with colorful models of molecules.

Molecule Math

Omar works with many different kinds of molecules—groups of atoms bound together. Every molecule can be represented by a formula. Take water. Its chemical formula, H_2O, means each water molecule contains two hydrogen (H) atoms, and one oxygen (O) atom, for a total of three atoms. Now count how many of each kind of atom are in the following molecules.

- Vinegar—$C_2H_4O_2$
- Glucose—$C_6H_{12}O_6$
- Chalk—$CaCO_3$

Find the formulas for five more molecules. Then challenge a classmate to figure out how many of each kind of atom there are in the molecules you selected.

A Drop of Science

How can you show the 3-D shape of molecules using models? Start with something round, such as gumdrops, to represent atoms. Pick a different color for each of the following atoms—carbon, hydrogen, and oxygen. Use toothpicks to represent the bonds that link atoms. Then build the three molecules below.

1. Methane—CH_4
   One carbon atom and four hydrogen atoms

2. Carbon dioxide—CO_2
   One carbon atom and two oxygen atoms

3. Water—H_2O
   Two hydrogen atoms and one oxygen atom

Check out your answers on page 36.

Organic Chemist

"I majored in chemistry because I wanted to find out what was in the world around me."

Dorian Canelas

Duke University

Fizz-ics? No, Chemistry!

Next time you slide a fried egg from pan to plate, think of Dori Canelas! She helped find a greener way to make the nonstick plastic that lines the pan. Making that slippery plastic used to require the help of lots of harmful chemicals called solvents. Those solvents include chlorofluorocarbons, benzene, carbon tetrachloride, and other chemicals that are as nasty as they sound. So Dori and her colleagues found a safer and surprising substitute solvent—carbon dioxide, or CO_2. It's the gas you exhale and the gas that puts the fizz in your soda pop.

Squeeze for Luck

Dori and her colleagues already knew CO_2 worked as a solvent—companies use it to dissolve the caffeine in coffee beans. But they weren't sure if CO_2 could replace the harmful chemicals used in making nonstick plastic. So they gave it a try. It worked! "We squeeze down CO_2 gas into liquid, and then use it to help make plastic," Dori explains. "Our research produced a real process—not just something that was filed away in a book somewhere," she says. *Egg*-cellent!

Chloro What?

As a young girl, Dori loved to puzzle over ingredient labels. "I didn't even know what half the words were—until I took high school chemistry," Dori says. "That's what got me into chemistry— an interest in finding out what was in the world around me."

Dori still plays volleyball, like she did in college. She also teaches and works in her lab.

An organic chemist

studies the chemistry of organic compounds—compounds that contain carbon and hydrogen. Dori researches using nontoxic solvents to make plastics. Other **organic chemists**

* alter natural carbon compounds for use as new medicines.
* identify the chemical compounds insect pests use to attract mates.
* study proteins, the building blocks of cells.
* design reactions that turn vegetable oils into margarine.
* create new Earth-friendly pesticides to protect crops.

"It's not just sitting at a desk," Dori says of her life.

About You

Dori's dad told her to find something she loves to do. Write an entry in your About Me Journal describing something you love to do. How could you turn this into a future career?

H_2O Ha, Ha

Q. H_2O is the formula for water. What's the formula for ice?

A. H_2O cubed!

Chemistry Is a Gas!

With your teacher's permission and help, try this experiment. Don't forget to wear safety goggles.

* Pour 60 milliliters (¼ cup) of distilled water into a plastic soda bottle—either outside or over a sink. Next add the same amount of white vinegar.
* Use a funnel to fill a deflated balloon halfway with baking soda.
* Let the balloon flop to the side as you stretch the neck of the balloon tightly over the top of the bottle. Then hold up the balloon so the baking soda falls into the liquid. Be sure to hold onto the balloon where it's stretched over the bottleneck!

What happens to the balloon? Why? Develop a hypothesis. Stumped? Read page 22 again.

Check out your answers on page 36.

MARTIN WOLF

Seventh Generation, Inc.

Faucet Fiddler

As a teenager, Martin Wolf made a discovery. If he adjusted the hot and cold faucets in the shower just right, it reduced the flow of water and kept the water perfectly warm. Hey! Martin had found a way to save water *and* save energy. Years later, Martin is still putting the green in clean.

Cleaner Cleaners

Today, Martin creates laundry detergents, dish soaps, and other household cleaners. Each product is easy on the environment—and safe to use. "Other companies say it's okay to use hazardous materials as long as people use them properly," Martin says. "My feeling is, why not eliminate the hazard from the start?" He uses ingredients that are sustainable—made from renewable plant sources. That ensures each ingredient breaks down in the environment to form substances that are harmless—or even helpful, such as fertilizer to grow more plants. "You can create a cycle just like nature does," Martin says.

Friend of Fish

Martin works with lawmakers to ban the use of phosphate in dishwashing detergent. When phosphate drains into lakes, rivers, or streams, it causes tiny organisms called phytoplankton to multiply by the zillions in the water. They soak up the oxygen that fish and other organisms need to stay alive.

"Why not do things in a way that minimizes our impact on the environment?"

A product technologist

creates new or improved products. Martin uses green chemistry to create new kinds of eco-friendly cleaning products. Other eco-friendly **product technologists** make

* carpets that don't give off toxic fumes.
* plastic containers free of heavy metals.
* car parts that can be easily recycled.

Shine On

Many metal cleaners contain harmful chemicals. So, whip up your own and test them. Ask your family for tarnished metal items that are made of copper, brass, or stainless steel. As a class, organize the collected items by the type of metal. Then start scrubbing.

- Brass and copper—Add lemon juice to baking soda and make a paste. With a soft cloth, rub the paste on half of an item. Rinse and dry.
- Stainless steel—Dip a soft cloth in white or cider vinegar. Wipe half of the surface.

As a class, compare the clean half and the tarnished half of each metal item. Did your green cleaner do the job?

Keep It Green

From his office in Vermont, Martin can see beautiful Lake Champlain. What would pollution do to it? Brainstorm with a partner how pollution might change beautiful places you've seen. Choose one place and share with the class where and what it is, and how pollution could change it.

Is It 4 U?

At his job, Martin enjoys

- using chemistry to make environmentally safe products.
- working with a team.
- educating consumers about the importance of eco-friendly products.

Discuss with a partner which parts of Martin's job interest you. What do you think would make you a good green chemist? Why?

NNEKA BREAUX
Dow AgroSciences

Questions, Questions Everywhere

When she was young, Nneka Breaux wasn't sure if she was going to be a lawyer, a teacher, or a scientist. One thing she was sure of—she had a lot of questions. "I found out that asking questions was just the beginning. The coolest part is you can find the answers if you actually *do* something," Nneka says. As a chemist, that's exactly what she does today.

A Plant-astic Job

Nneka starts with an idea, such as finding new ways to fight insect pests. That might get her tinkering with a naturally occurring chemical compound. She makes sure it targets *only* harmful insects—and not helpful insects, such as bees or ladybugs. Nneka then makes the new compound and sends it to her company's greenhouse for testing. "I get to see the big picture in science—I actually see things move from an idea to something that works," Nneka says. The crop protection products Nneka helps create replace older, more toxic chemicals used in orchards. That lets farmers grow healthy fruit in healthy ways. So the next time you bite into a juicy peach, thank chemists like Nneka who ask questions—and look for answers.

Here, There, Every Year a Fair

Nneka competed in science fairs from first grade through ninth grade. A magnetic compass was just one of her projects—you might say it helped Nneka find her way in science!

"I can design molecules, and then send them to the greenhouse to be applied to the plant and see if they work the way we planned."

A research chemist

studies chemical compounds and their reactions to create new compounds and products. Nneka designs environmentally friendly pesticides. Other **research chemists**

✳ create new products from plant compounds.

✳ design medicines with fewer side effects.

✳ develop aroma chemicals used in perfumes.

✳ explore how food ingredients work together.

It Never Hurts to Ask

Nneka thinks "the coolest part" of coming up with a question is figuring out the answer.

- With a partner, think of a question that hasn't been answered.
- How would you go about investigating an answer to your question?
- What training and career would help you answer your question?

Find out if someone is currently looking for an answer to your question. Write a brief description of that person's job.

Chirping Thermometer

A chirping cricket makes a pretty good thermometer. Really, it does! A formula for the relationship between the number of cricket chirps and air temperature was published in 1898. Here's that formula.

$$°F = 40 + n$$

°F stands for the temperature in degrees Fahrenheit and n is the number of cricket chirps per 15 seconds. First, find the temperature in degrees Fahrenheit when crickets chirp at a rate of 128 chirps per *minute*. Next, convert your answer from degrees Fahrenheit to degrees Celsius. Here's the formula.

$$°C = \frac{5}{9} \, (°F - 32)$$

What's the cricket weather report?

Check out your answers on page 36.

KAICHANG LI
Oregon State University

Soy Far, Soy Good

Kaichang's glue is more environmentally friendly than older glues made from formaldehyde, which is toxic and can cause allergies and asthma. A plywood company used Kaichang's soybean formula to replace about 20 million kilograms (about 44 million pounds) of formaldehyde in one year alone!

The Muscle of Mussels

Have you ever looked around you and picked up a totally cool idea? That's what happened to Kaichang Li while doing something he loves—exploring the outdoors. Walking along the Oregon coast, he noticed how small shellfish called mussels stuck to the rocks. Why didn't they wash away when strong waves smashed into them? Kaichang took some of the mussels back to his lab to study them.

Sticking with It

The little mussels keep a tight grip on the rocks by using a strong but flexible thread-like protein they secrete. Kaichang identified the protein, and then looked for ways to make more of it. His source? Plentiful and cheap soybeans. Kaichang chemically alters a protein found in soybean flour to make glue that sticks things together as firmly as a mussel grips a rock! The glue is strong enough to use in making desks and chairs. And it's safe enough to hold together baby cribs. Kaichang's formula is so simple and safe, he can whip up a batch using a kitchen mixer!

A wood chemist finds

news ways to use and manufacture wood products. Kaichang invents environmentally friendly glues used in plywood. Other **wood chemists**

* study diseases that affect trees.
* reduce the waste in making paper.
* develop treatments that protect wood from decay.
* discover chemicals in wood that are useful as medicines.

Soy What?

Soybeans are packed with protein, and easy to grow. Their uses in food products seem endless—check the ingredients on packaged food. Soybeans are also rich in oil that's used to make products such as these.

Candles	Foam cushions
Soaps	Crayons
Plastics	Biofuels
Inks	Clothing

Form a small group and choose one product from the list to research. Write a group report including

* how soy is used in the product.
* what the soy substitute replaces.
* a sketch of what the product looks like.

Then, as a class, combine reports and create a consumer booklet about the many uses of soy.

About You

Kaichang's discovery started with something that made him wonder. In your About Me Journal, write about something that makes you wonder. And keep wondering—you never know where it might take you.

Sticky Numbers

One company used Kaichang's soybean glue to replace 20 million kilograms (about 44 million pounds) of formaldehyde in one year. Both numbers are written in short word form. Now write both numbers in standard form.

How many kilograms and pounds of formaldehyde would be replaced in two years? Write those numbers in short word form, in standard form, and in expanded form.

Check out your answers on page 36.

About Me

The more you know about yourself, the better you'll be able to plan your future. Start an **About Me Journal** so you can investigate your interests, and scout out your skills and strengths.

Record the date in your journal. Then copy each of the 15 statements below, and write down your responses. Revisit your journal a few times a year to find out how you've changed and grown.

1. *These are things I'd like to do someday.*
   Choose from this list, or create your own.

   - Study chemistry and the environment
   - Protect and improve community health
   - Design Earth-friendly chemical processes
   - Give lectures
   - Write books
   - Develop and test new materials
   - Invent cleaner, safer, better products
   - Create ways to reduce waste
   - Teach environmentally friendly science
   - Find alternatives to harmful chemicals
   - Develop ways to fight pollution

2. *These would be part of the perfect job.*
   Choose from this list, or create your own.

   - Working outdoors
   - Building things
   - Writing
   - Working on teams
   - Observing
   - Traveling
   - Researching
   - Drawing
   - Solving problems
   - Leading others

3. *These are things that interest me.*
Here are some of the interests that people in this book had when they were young. They might inspire some ideas for your journal.

- Dancing
- Exploring salt marshes
- Watching birds
- Writing stories
- Experimenting with a chemistry set
- Investigating the natural world
- Exploring the outdoors
- Hiking
- Researching ingredients in products
- Competing in science fairs
- Sailing
- Skiing

4. *These are my favorite subjects in school.*

5. *These are my favorite places to go on field trips.*

6. *These are things I like to investigate in my free time.*

7. *When I work on teams, I like to do this kind of work.*

8. *When I work alone, I like to do this kind of work.*

9. *These are my strengths—in and out of school.*

10. *These things are important to me—in and out of school.*

11. *These are three activities I like to do.*

12. *These are three activities I don't like to do.*

13. *These are three people I admire.*

14. *If I could invite a special guest to school for the day, this is who I'd choose, and why.*

15. *This is my dream career.*

Green Chemistry

Which ∧ career is 4 U?

What do you need to do to get there? Do some research and ask some questions. Then, take your ideas about your future—plus inspiration from scientists you've read about—and have a blast mapping out your goals.

On paper or poster board, map your plan. Draw three columns labeled **Middle School, High School,** and **College.** Then draw three rows labeled **Classes, Electives,** and **Other Activities.** Now, fill in your future.

Don't hold back—reach for the stars!

Air Quality Chemist

Entomologist

Environmental Quality Director

Public Health Expert

Chemical Engineer

Research Chemist

Toxicologist

Recycling Chemist

Chemistry Teacher

Green Chemistry
Professor

Biochemist

Tropical
Ecologist

Environmental Reporter

Product
Technologist

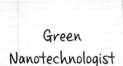
Polymer Chemist

Agricultural
Chemist

Materials
Scientist

Organic
Chemist

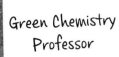
Environmental Health
Scientist

Green
Nanotechnologist

Wood
Chemist

Mircrobiologist

Manufacturing
Chemist

Biomedical
Engineer

Renewable Energy
Engineer

Environmental
Engineer

Glossary

biofuel (n.) A fuel made from renewable resources such as plants or wastes. It replaces or reduces the use of fossil fuels such as gasoline. (pp. 10, 11, 15, 29)

carbon dioxide (n.) (CO_2) A colorless gas. It is naturally present in small amounts in the atmosphere. The increase of carbon dioxide in the atmosphere, from the burning of fossil fuels and the destruction of vast areas of forest, is the cause of global warming. Carbon dioxide dissolves in water to form carbonic acid, is formed during animal respiration and the decay of animal or plant matter, is absorbed from the air by plants in photosynthesis, and is used to carbonate beverages. (pp. 20, 21, 22, 23)

Celsius (°C) (adj.) Temperature scale in which the freezing point of water is 0 degrees and the boiling point of water is 100 degrees. (p. 27)

chemical reaction (n.) A process where one or more substances are changed into new substances with different properties (pp. 10, 11, 23)

chlorofluorocarbons (CFCs) (n.) A synthetic chemical used in aerosols as a propellant, in refrigerators and in air conditioners as a coolant, and in foam packaging. Although CFCs are chemically inert, they can remain in Earth's atmosphere for more than 100 years. CFCs destroy the ozone layer. (p. 22)

element (n.) Any substance that exists in its purest chemical form. (p. 11)

engineering (n.) The application of science, math, and technology to design materials, structures, products, and systems. (pp. 9, 18)

Fahrenheit (°F) (n.) A temperature scale. On this scale, the freezing point of water is 32 degrees and the boiling point of water is 212 degrees, and the interval between is divided into 180 equal parts. (p. 27)

fossil fuels (n.) Nonrenewable energy resources such as coal, oil, and natural gas that are formed from the compression of plant and animal remains over hundreds of millions of years. (p. 18)

green chemistry (n.) The design of chemical products and processes that reduce or eliminate hazardous substances. Green chemistry provides many benefits, including less waste, less costly cleanup, lower energy use, and safer products for people and the environment. (pp. 9, 10, 11, 14, 15, 16, 25)

greenhouse gases (n.) Gases such as carbon dioxide, water vapor, and methane that absorb infrared radiation or heat. These gases absorb some of the infrared radiation trying to escape Earth instead of letting it pass through the atmosphere, resulting in a greenhouse effect. (p. 20)

methane (n.) A colorless, odorless, and highly flammable gas that results from the decomposition of organic matter or the carbonization of coal. (pp. 20, 21)

nanotechnology (n.) The technology of building machines, such as electric motors, and eventually whole robots, on a very small scale—the scale of atoms and molecules. Nanotechnology devices are typically only a few nanometers in size. (p. 16)

solar panels (n.) A device, made of solar cells, used to convert the energy of sunlight into electricity. (p. 16)

solvents (n.) A liquid in which a substance will dissolve to form a solution. (pp. 6, 21, 22, 23)

Index

CHECK OUT YOUR ANSWERS

MATERIALS SCIENTIST, page 21

Molecule Math
Vinegar, $C_2H_4O_2$
C (carbon) 2, H (hydrogen) 4, O (oxygen) 2—8 atoms total
Glucose, $C_6H_{12}O_6$
C (carbon) 6, H (hydrogen) 12, O (oxygen) 6—24 atoms total
Chalk, $CaCO_3$
Ca (calcium) 1, C (carbon) 1, O (oxygen) 3—5 atoms total

ORGANIC CHEMIST, page 23

Chemistry is a Gas!
Inside the plastic bottle, a chemical reaction takes place that causes the liquid to bubble and fizz. The bubbles are carbon dioxide—a gas formed when an acid (vinegar, or weak acetic acid) and a base (baking soda, or sodium bicarbonate) chemically react. This is the gas that fills the balloon. It's the same gas that gives soda pop its fizz!

RESEARCH CHEMIST, page 27

Chirping Thermometer
The temperature is 72° Fahrenheit.
n = chirps per 15 seconds = 32

$$32 = \frac{128 \text{ chirps}}{\text{minute}} \times \frac{1 \text{ minute}}{60 \text{ seconds}} \times 15 \text{ seconds}$$

72 degrees Fahrenheit = 40 + 32 (chirps in 15 seconds)

Converted to Celsius, the temperature is 22.2°.

$$22.2 = \frac{5}{9} \times (72 - 32)$$

WOOD CHEMIST, page 29

Sticky Numbers
The amounts written in standard form are 20,000,000 kilograms formaldehyde in two years and 44,000,000 pounds formaldehyde in two years.

In two years,
20,000,000 kilograms × 2 years =
    40 million kilograms (short word form)
    40,000,000 kilograms (standard form)
    $4 \times 10,000,000$ (expanded form)

44,000,000 pounds × 2 years =
    88 million pounds (short word form)
    88,000,000 pounds (standard form)
    $8 \times 10,000,000 + 8 \times 1,000,000$ (expanded form)

IMAGE CREDITS

Fernando Audibert: Cover. Erin Hunter: pp 2-3 and pp 30-31 background, p. 17. Yale University: p. 2 (Anastas), p. 14 top. University of Oregon: p. 2 (Hutchison), p. 16 top. University of North Texas: p. 3 (D'Souza). © 2009 UC Regents: p. 3 (Yaghi), p. 20 bottom, p. 21. Oregon State University: p. 3 (Li), p. 28. Exxon Mobil: p. 4. NASA: p. 5. Rob Landis: p. 10. Heinz Awards Photo / Jim Harrison photographer: p. 14 bottom. Oxford University Press: p. 15. Laurie Monico: p. 16 bottom. Photo by University of North Texas/URCM Photography: p. 18 top. Scott R. Bauer, BAUER Photographics, Inc.: p. 18 bottom. OMNI Ventures: p. 19. Clara Lam: p. 30. California State University, Fullerton: p. 32 bottom left. U.S. Department of Energy: p. 33 top left. William A. Cotton/Colorado State: p. 33 bottom left. Courtesy Callie McConnell, Desert Research Institute: p. 33 top right. Natural Resources Defense Council: p. 33 bottom right.